Humpback Whale Migration

By Susan H. Gray

21st Century Junior Library

Cherry Lake Press

Published in the United States of America by Cherry Lake Publishing Group
Ann Arbor, Michigan
www.cherrylakepublishing.com

Reading Adviser: Marla Conn, MS, Ed., Literacy specialist, Read-Ability, Inc.
Content Adviser: Sheila K. Schueller, Ph.D., Lecturer and Academic Program Specialist, University of Michigan School for Environment & Sustainability
Photo Credits: ©Tomas Kotouc/Shutterstock.com, cover; ©Earth theater/Shutterstock.com, 4; ©Pascale Gueret/Shutterstock.com, 6; ©Jay Ondreicka/Shutterstock.com, 8; ©Shane Gross/Shutterstock.com, 10; ©Imagine Earth Photography/Shutterstock.com, 12; ©RLS Photo/Shutterstock.com, 14; ©WorldFoto/Alamy Stock Photo, 16; ©18 Nico Faramaz/Shutterstock.com; ©Joe West/Shutterstock.com, 20

Cherry Lake Press is an imprint of Cherry Lake Publishing Group.

Library of Congress Cataloging-in-Publication Data

Names: Gray, Susan Heinrichs, author.
Title: Humpback whale migration / by Susan H. Gray.
Description: Ann Arbor, Michigan: Cherry Lake Publishing, [2021] | Series: Marvelous migrations | Includes index. | Audience: Grades 2-3
Identifiers: LCCN 2020002734 (print) | LCCN 2020002735 (ebook) | ISBN 9781534168596 (hardcover) | ISBN 9781534170278 (paperback) | ISBN 9781534172111 (pdf) | ISBN 9781534173958 (ebook)
Subjects: LCSH: Humpback whale—Migration—Juvenile literature.
Classification: LCC QL737.C424 G73 2021 (print) | LCC QL737.C424 (ebook) | DDC 599.5/251568–dc23
LC record available at https://lccn.loc.gov/2020002734
LC ebook record available at https://lccn.loc.gov/2020002735

Cherry Lake Publishing Group would like to acknowledge the work of the Partnership for 21st Century Learning, a Network of Battelle for Kids. Please visit http://www.battelleforkids.org/networks/p21 for more information.

Printed in the United States of America
Corporate Graphics

CONTENTS

The humpback whale has distinct bumps on its head.

Bulking Up

Ahh! Summer in the **Arctic**. The sea is loaded with **krill** and fish. The biggest whales eat 3,000 pounds (1,361 kilograms) every day. These giant **mammals** weigh around 80,000 pounds (36,287 kg)! They are **bulking** up for the long trip ahead.

Winter is coming, and humpbacks are preparing for their **migration** to warmer seas. Pacific whales from Russia to Alaska

No two whales have the same tail.

are chowing down. Atlantic whales from Maine to Norway are fattening up. Their layer of **blubber** is growing thicker. It will provide their energy for months. Humpbacks will spend the winter in warm waters near the **equator**. Once they leave the Arctic, they will eat little or nothing until they return.

Look!

The tail of a humpback is called a fluke. Scientists often identify individual humpbacks by their flukes. Look at pictures of some whales' flukes. What differences, even small ones, do you see?

Most whales have one blowhole, but humpback whales have two!

Getting Warmer

Humpbacks live together in large groups, often with thousands of members. When they migrate, they travel in smaller groups called pods. Usually, moms with calves are the first

Make a Guess!

One group of humpbacks lives in the Arabian Sea. This is a small group with fewer than 100 whales. They never migrate. Why do you think they just stay there?

Female whales are larger than males.

to leave. As they travel, mothers often pat their young with their fins. Young humpbacks leave next, followed by adult males. The last to abandon the feeding area are the females that will soon give birth. They need to eat as much as they can before they go.

In the Northern **Hemisphere**, Atlantic humpbacks swim toward Africa or islands near Florida. Pacific whales head to Mexico, Central America, Hawaii, or islands near Japan. Wherever they go for the winter, it is always toward warmer water.

This is where mothers give birth to their 3,000-pound (1,361 kg) babies. These

A humpback whale's pregnancy lasts 11 months!

newborns are a little clumsy at first. But their moms stay close by, often swimming right beneath them.

This area is also where **courtship** takes place. In time, a pair will form and mate. The female will have her baby here about this time next year.

Ask Questions!

Humpbacks are known for their beautiful "songs." These are long, slow notes sung only by the males. They may help in attracting mates. But they might also have other uses. Look online to see what different experts say.

Krill travel in groups of thousands!

North and South

As winter ends, the humpbacks begin moving north. They head to the same Arctic feeding areas they left earlier. The water there is full of krill and small fish. By now, humpbacks are *really hungry*. When they arrive, the whales are happy to feast again.

The Northern Hemisphere whales have left the Arctic and returned. But what

Humpback whales swim in circles and blow
bubbles to confuse and capture prey. This is called bubble-net feeding.

about the humpbacks in the Southern Hemisphere? They spend their summers close to **Antarctica**. Like their northern cousins, they feed and fatten up. Then, they swim to warmer waters for the winter. These whales migrate to South America, Africa, and Australia. After calves are born and pairs mate, back to Antarctica they go!

Think!

Some humpbacks travel from Antarctica to the islands of Tonga. The whale visits are good for business in Tonga. What kinds of businesses might benefit?

Humpback whales will leap out of the water.

A Whale of a Trip

Every year, humpbacks travel from feeding areas to breeding areas and back. For some, the round trip can be more than 10,000 miles (16,093 kilometers)! It is the longest known migration of any mammal.

The biggest whales gain thousands of pounds before the trip. They lose it all by the time they return. Babies are born 15 feet (4.6 meters) long and grow rapidly. They

Humpback whales used to be endangered, but now their populations are growing.

complete their first full migration about the time they turn 1 year old.

Scientists want to know more about humpbacks. What do their songs really mean? What makes them start a migration? Can underwater videos tell us more about their size and behavior? How many humpbacks are there, anyway? And what's the best way to count them? There's so much more to learn about these giants of the sea.

GLOSSARY

Antarctica (ant-ARK-tih-kuh) the landmass around the South Pole

Arctic (AHRK-tik) pertaining to the area around the North Pole

blubber (BLUHB-ur) a layer of fat in sea mammals such as whales, seals, and walruses

bulking (BUHLK-ing) gaining weight

courtship (KORT-ship) the behavior of animals in order to find a mate

equator (ih-KWAY-tur) an imaginary line around the Earth halfway between the North Pole and South Pole

hemisphere (HEM-ih-sfeer) half of the globe

krill (KRIL) small, shrimplike animals that live in the ocean

mammals (MAM-ulz) animals that have hair and give birth to live young

migration (mye-GRAY-shuhn) movement of animals from one region to another and back again

FIND OUT MORE

BOOKS

Gunderson, Megan M. *Humpback Whales*. Edina, MN: ABDO, 2011.

Leaf, Christina. *Humpback Whales*. Minneapolis, MN: Bellwether Media, 2017.

Marsh, Laura. *Great Migrations*: *Whales*. Washington, DC: National Geographic Children's Books, 2010.

WEBSITES

ActiveWild—Humpback Whale Facts
https://www.activewild.com/humpback-whale-facts
Read more about this whale and learn about its songs, body, and behavior.

YouTube—Whales' Bubble Net Fishing
https://www.youtube.com/watch?v=Q8iDcLTD9wQ
Watch how the humpbacks blow bubbles to confuse, trap, and catch fish.

INDEX

ABOUT THE AUTHOR

Susan H. Gray has a master's degree in zoology. She has written more than 170 reference books for children and especially loves writing about animals. Susan lives in Cabot, Arkansas, with her husband, Michael, and many pets.